From Day One to Success

A Step-by-Step Guide for Scrum Masters Joining a New Organization

KAREN FOMAFUNG

ISBN: 979-8-850451-19-6

Cover design by: BeingAgile Consulting
Edited by: BeingAgile Consulting LLC
Some images were taken from pexel.com's public domain.
Library of Congress Control Number: 2018675309
Printed in the United States of America

DEDICATION

I dedicate this book to my BeingAgile Consulting Rocket Team - Level 02 Scrum Master Mentorship community, who consistently asked for my guidance on how to hit the ground running after landing their first job offer as a Scrum Master, leading to my conception of the idea From Day One to Success.

Please report any mistake to info@beingagileconsulting.com
First Edition - July 2023

CONTENTS

ACKNOWLEDGMENTS

I would like to express my deepest gratitude and appreciation to my family, friends, colleagues, and students who have contributed from the conception to the delivery of this book, "From Day One to Success: A Step-by-Step Guide for Scrum Masters Joining a New Organization. Your wisdom, experience, and constructive feedback have significantly enriched its content.

First and foremost, I want to thank my beautiful kids, Royal Fomafung and Heavenly-Peace Fomafung, who are why I was determined to complete this book. I always teach them the importance of stop-starting and start-finishing. They held me down on this project, and because I wanted to lead by example, I kept going until DONE.

I want to thank my BeingAgile Consulting community; Professionals, mentees, students, and all the dedicated Scrum Masters who tirelessly work to foster Business Agility and enable organizational success. Your passion, courage, and commitment to succeed with Agile inspired me to write this book and share my experience. I am immensely grateful to the professionals and organizations who graciously shared their real-world experiences and case studies, allowing me to present practical examples and lessons learned.

Your contributions have added depth and authenticity to the guidance offered within these pages. Lastly, a special thank you goes to my family for their unwavering love, patience, and understanding. Your support and belief in me have been my constant motivation. To all who have played a part, big or small, in the creation and publication of this book, please accept my heartfelt thanks. Your contributions have been invaluable, and I sincerely appreciate your involvement.

Karen Fomafung

VALUE PROPOSITION

This inspiring and comprehensive guidebook takes you through the crucial first 90 days as a Scrum Master and beyond, providing practical step-by-step strategies, real-world scenarios, tools, and insights to thrive in your new role. This book covers everything from preparing to join your organization, building relationships, and guiding Agile adoption to facilitating effective meetings, managing conflicts, and leaving a positive legacy. With authenticity, humor, and actionable advice, this guide empowers Scrum Masters to navigate challenges, foster collaboration, and achieve continuous improvement.

Whether you're a beginner or experienced Scrum Master, starting in a new organization, assigned to a new team, moving to another part of a company, or seeking to enhance your skills in an existing role, this step-by-step guide will equip you with the knowledge and tools to excel. This book is your roadmap to success in the dynamic world of Agile in your new organization.

PREFACE

Welcome to "From Day One to Success: A Step-by-Step Guide for Scrum Masters Joining a New Organization." This book is dedicated to all the passionate Scrum Masters embarking on a new chapter in their careers, ready to make a significant impact in their new organizations.

Joining a new organization as a Scrum Master can be exciting and challenging. The first 90 days play a critical role in setting the foundation for your success and shaping the trajectory of your journey. During this time, you have the opportunity to establish strong relationships, understand the organizational culture, and positively impact the teams you serve.

This guide is designed to be your trusted companion, providing practical insights, strategies, and a step-by-step roadmap to navigate the intricacies of your new role. Each chapter is carefully crafted to address specific aspects of your journey, from preparing to join the organization and making a memorable entrance to fostering a positive team culture, managing challenges, and leaving a lasting legacy.

As you embark on this transformative journey, remember you are not alone. This book is a culmination of my years of experience as a Scrum Master and Agile practitioner, combined with the collective wisdom and lessons learned from countless Scrum Masters who have navigated similar paths. I hope this guide will empower you with the knowledge, confidence, and inspiration to thrive in your new role, make a meaningful impact, and achieve lasting success.

Remember, your journey starts from day one, and this book will be your trusted guide. Embrace the challenges, cultivate your skills, and transform organizations with your Agile leadership. Get ready to embark on a remarkable adventure as a Scrum Master, and may your path be filled with continuous growth, fulfillment, and success.

Wishing you a rewarding and transformative journey.

Follow us on BeingAgile Consulting.

Website: www.beingagileconsulting.com
Call/WhatsApp: (+1) 612-441-9214

INTRODUCTION

From Day One to Success: A Step-by-Step Guide for Scrum Masters Joining a New Organization.

Welcome to the Scrum Master's first 90 days and beyond! I am thrilled to have the opportunity to share my real-world experiences and insights with you as we embark on this exciting evolution together. Over 13 years as an Agile Practitioner, I have worked with various organizations, mentoring over 3000 Scrum Masters. Through this book, I aim to provide a step-by-step guide that will empower you to confidently navigate your first 90 days and set the stage for ongoing success. Please note that there is no one-size-fits-all regarding Business Agility. Organizations adopt Agile differently based on structure, challenges, systems, etc. The insights I will share are based on my experiences as a Scrum Master, a mentor to thousands of Scrum Masters, and an Agile Coach. So, the purpose of this book is to serve as a guide and not a standard rule.

Chapter by chapter, we will explore the critical aspects of being a Scrum Master in a new organization. From preparation and making a memorable entrance on your first day to leaving a positive and sustainable legacy if you choose to transition out, we will cover the essential things you need to know to thrive in your role. Before diving into the specifics, let's take a moment to understand why the first 90 days are crucial.

I. The Importance of the First 90 Days

The first 90 days of joining a new organization as a Scrum Master holds immense significance in shaping the trajectory of your journey. This crucial period sets the tone for your relationships, establishes your credibility, and determines your effectiveness as a facilitator of Agile practices. During this time, you can build trust, gain insights into the team dynamics, and immerse yourself in the organization's culture. Investing your energy and focus in these initial days can lay a solid foundation for long-term success. Through your proactive engagement, open mindset, and willingness to learn and adapt, you can make a lasting impact, foster collaboration, and steer your team toward achieving Agile excellence. Embrace the challenges, seize the opportunities, and approach each day with a sense of purpose, for within these first 90 days, you have the power to shape your future as a remarkable Scrum Master.

II.   Setting the Tone: Being Authentic and Building Rapport

Authenticity is vital to building rapport and connecting with your team. In Chapter 2, we will delve into the art of introducing yourself with

confidence and authenticity. We will explore how to break the ice, engage in meaningful conversations, and create a positive atmosphere that fosters collaboration and trust. Remember, being yourself is the best way to build genuine connections and create a supportive environment.

III.   Using Humor as a Tool for Connection

As a Scrum Master, part of your job is to be the light in the room. Laughter is a universal language that can break down barriers, ease tension, and strengthen relationships. This book will explore humor's power as a connection tool. I will share practical examples of how humor can be infused into your interactions, meetings, and daily routines as a Scrum Master. Embracing a lighthearted approach can foster creativity, resilience, and positive team culture.

As we progress through each chapter, I will draw upon my own experiences and share authentic stories, anecdotes, and case studies from the field. Learning should be enjoyable, so I have infused humor and witty insights

throughout the book. At the end of each chapter, you will find exercises, reflection questions, and actionable tips to enhance your learning experience and help you apply the concepts to your evolution.

So, whether you are starting as a Scrum Master in a new organization, assigned to a new team, moving to another part of a company, or seeking to enhance your skills in a current role, this step-by-step guide will equip you with the knowledge and tools to excel. Get ready to embrace the adventure, navigate challenges with resilience, and establish a positive legacy as a Scrum Master.

Now, let's embark on this exciting evolution together, from preparing to start and making a memorable entrance on day one to lasting success!

CHAPTER 1
PREPARING TO JOIN YOUR NEW ORGANIZATION

This chapter will explore the essential tools and techniques every Scrum Master should focus on before their official first day to help them prepare to start right in their new organization. You can lay a strong foundation and confidently navigate your new role by familiarizing yourself with the organization's culture, people, and tools. Let's get started!

I. Researching and Understanding Your Company's Culture and People

Researching and understanding the culture and people of a company before joining as a new Scrum Master is paramount. It allows you to gain invaluable insights into the organization's values, norms, and working dynamics, enabling you to navigate the new environment more effectively.

To embark on this crucial endeavor, start immersing yourself in the company's online presence. Explore their website, particularly their mission, vision, and core values. These foundational elements serve as guideposts, offering a glimpse into the company's overarching goals and principles. Delve into their social media platforms, blog posts, and any available press releases to gather information about their culture, recent initiatives, and how they present themselves to the public.

While online research provides a solid starting point, nothing compares to firsthand experiences and conversations. Reach out to current and former employees through professional networking platforms, industry events, or personal connections. Engage in meaningful discussions to gain insights into the work environment, team dynamics, leadership styles, and the company's overall atmosphere. Actively listen to their perspectives and stories, focusing on positive aspects and areas that may warrant further exploration.

Scenario: It was an exciting time when I accepted a position at a renowned software development company, and I wanted to make a positive impact right from the start.

Before my official start date, I took the initiative to immerse myself in the company's culture. I delved into their website and social media platforms, carefully studying their mission, vision, and core values. This gave me a

glimpse into the organization's priorities and helped align my mindset with their overarching goals.

But I didn't stop there. I reached out to my professional network to connect with current and former employees of the company. Through insightful coffee chats, video calls, and LinkedIn messages, I engaged in candid conversations about the work environment, team dynamics, and management style. These interactions provided invaluable perspectives and allowed me to anticipate potential challenges and prepare strategies to foster a positive team culture right from day one.

To better understand the company's Agile ways of working, I sought out a fellow Scrum Master with experience within the organization. During an informational interview, they shared insights on the team's Agile maturity level, preferred tools and techniques, and unique aspects of their Agile implementation. Armed with this knowledge, I researched and familiarized myself with the company's specific Agile frameworks and tools, ensuring I was well-prepared to hit the ground running.

I also made it a point to attend industry events and join online communities dedicated to Agile and Scrum. Actively participating in discussions, asking questions, and sharing my experiences allowed me to gain a broader perspective on current trends and challenges in Business Agility. I learned from professionals across various organizations, which helped me refine my approach, gather best practices, and avoid potential pitfalls.

By investing time and effort in researching and understanding the culture and people of the company even before my start date, I demonstrated my commitment to success as a Scrum Master. Armed with some understanding of the company's values, norms, and Agile practices, I was able to build healthy relationships, adapt my approach accordingly, and foster a positive and productive work environment right from the beginning.

II.   Researching and Understanding Your Company's Tool Stack

As a new Scrum Master, it's crucial to research and understand the tools and Agile Project management approach used in your new organization before your employment start date. By exploring these tools in advance, you can better understand their functionalities, benefits, and how they support the Agile ways of working. Here are three common categories of tools organizations use, structured according to usability and purpose:

- **Project Management and Issues Tracking Tools:** Platforms like Jira, Trello, Azure DevOps, Rally or Monday.com, etc., are widely used for managing Agile projects. These tools help teams create backlogs, plan sprints, track tasks, and visualize progress. They provide a structured approach to managing the work progress, allowing Scrum Masters to facilitate effective planning and execution.

- **Communication and Collaboration Tools**: Tools such as Slack, Microsoft Teams, Zoom, WebEx, Google Suite - Gmail, Google Meet, and Google Hangout or Microsoft Outlook will be your primary means of facilitating effective team interactions. I have used them to conduct meetings, coordinate discussions, provide updates, and ensure continuous communication and collaboration among team members and stakeholders. They enable real-time messaging, video conferencing, virtual meetings, file sharing, and integration with other project management tools. They also ensure clear communication channels even when team members are geographically dispersed.

- **Documentation and Knowledge Sharing Tools**: Platforms like Confluence, Onedrive, or SharePoint allow teams to create and share documentation, process guidelines, group shared calendars, release calendars, and other project-related information. These tools promote transparency and ensure that essential project knowledge is easily accessible. As a Scrum Master, I have leveraged these tools to

document Agile processes, share meeting notes, create and prioritize Agile practice improvement backlog for myself and provide a central repository for critical project resources.

- **Virtual Whiteboard and Brainstorming Tools**: As a Scrum Master, you need a digital whiteboard and collaboration platform that enables virtual brainstorming, visual planning, and team collaboration, especially if your team is geographically dispersed. You need a tool that provides a canvas where teams can create diagrams, sticky notes, wireframes, and other visual artifacts. Tools like Mural, Zoom Whiteboard, Miro, etc., are good examples of virtual whiteboards used by most organizations. These tools help facilitate collaborative activities such as retrospectives, problem-solving, and workshop events. As a Scrum Master, I have leveraged these virtual whiteboards to capture ideas, gather feedback, and visually represent the team's work.

A. How to Prepare for Tool Usage

During the interview, I would always ask the interviewers for insights about their tool stack to proactively familiarize myself with the basics of these tools. Here is how you can prepare:

- **Learn the Basics**: Explore the features and functionalities of each tool by reviewing online tutorials, documentation, or video guides. Gain a basic understanding of how to navigate the interfaces from a Scrum Master Perspective, create and manage projects, report, and perform other everyday tasks.

- **Hands-on Practice:** Create a practice project or join a demo workspace such as a mentorship Community of Practice to gain hands-on experience with the tools. Experiment with different features, create sample backlogs, run simulated sprints, and explore collaboration capabilities. This hands-on experience will build your confidence and familiarity with the tools.

- **Seek Guidance:** Reach out to experienced Scrum Masters or Agile practitioners within your network and ask for their recommendations or best practices for using these tools. They can provide valuable insights and tips based on their own experiences.

B. Leveraging the Tools as a Scrum Master

As a Scrum Master, you will use these tools to facilitate Agile practices and support your team's success. Here are some common scenarios where you will need to utilize these tools:

- **Planning and Execution**: Agile project management tools will be crucial during sprint planning, backlog refinement, task tracking, and reporting. You'll need them to ensure that work items are well-defined, tasks are assigned, and progress is transparently visualized in collaboration with the team. As a Scrum Master, you will most likely facilitate the execution of these activities by the Developers and the Product Owner. To successfully facilitate the team's use of these tools, it is essential that you understand how to use them yourself.

- **Communication and Collaboration**: Communication and collaboration tools will be your primary means of facilitating effective team interactions. Use them to facilitate meetings, coordinate discussions, remove impediments, and ensure continuous communication between team members and stakeholders.

- **Documentation and Knowledge Sharing**: Documentation tools will support you in capturing and sharing important project information. You will need them to facilitate the maintenance of your team's process guidelines; meeting notes documentation, and other essential project resources. These tools foster transparency, enable knowledge sharing, and ensure that everyone has the necessary information at their fingertips.

In an organization where these tools were only being used after my joining, I seized the opportunity to introduce them to my teams and the organization, which was a massive contribution to their high-performance endeavor.

By proactively researching and familiarizing yourself with these tools before your first day, you'll be well-equipped to step into your role as a Scrum Master and hit the ground running. Embrace the learning process, seek guidance when needed, and be open to exploring additional tools that may align with your organization's specific needs. Your proactive approach will lay the foundation for a successful evolution as a Scrum Master in your new organization.

CHAPTER 2
DAY ONE: MAKING A MEMORABLE ENTRANCE

I.   Log in to the Organizational Systems and Complete Compliance Training

Ah, the daunting task of figuring out how to log in to your new work computer, navigating organizational systems, and completing compliance training. It may not be the most thrilling aspect of your first day, but it's an essential step towards smoothly integrating into the organization. Embrace the challenge with a sense of humor and an eagerness to master these systems. In my experience, on several occasions, an IT support contact was provided to me through the login process, from login into my computer for the first time, resetting my password, and logging in to other organizational systems as needed. Also, I have been in situations where I needed help to be made available to guide me through this daunting systems login process. So I had to call/email my manager to request login help. Usually, they will tell you who you will report to during the interview. Otherwise, be sure to ask. I have always found my managers very helpful during this phase. In situations where they were too busy to help me directly, they assigned another person to serve as my trail guide.

II.   Introducing Yourself with Confidence and Authenticity

Welcome to the beginning of an exciting evolution in a new organization! As someone who has spent over 13 years working with different organizations and mentoring over 3000 Scrum Masters, I understand the mixed emotions of starting fresh. It is crucial to approach your first day confidently and authentically, balancing between being a servant leader and a humble learner. How you introduce yourself plays a significant role in the perception of you. Remember, the first impression is mostly the lasting impression. Below are some examples of how I have introduced myself after joining a new organization as a Scrum Master

Scenario 01: Greetings, team! I'm thrilled to join this incredible team as your Scrum Master. With 13 years of experience, I've witnessed the power of Agile transformations and the endless possibilities it brings. But let me tell you a little secret—I'm just as eager to learn from each of you and contribute to our collective growth.

18

Scenario 02: Hello, team! I am Karen Fomafung, your trusty Scrum Master companion on this exciting evolution. Over the years, I've realized that the more I know, the more I realize there is to learn. So, let's embark on this adventure with confidence in our abilities and curiosity to explore the uncharted territories of team excellence.

III. Using Humor to Create a Positive Atmosphere

Laughter is a powerful tool that can transform any environment. As Scrum Masters, we can infuse humor into our interactions, creating a positive atmosphere that fosters collaboration, creativity, and enjoyment. Embrace your playful side and watch the team flourish. At the same time, be sure to strike a balance between humor and serious business.

IV. Breaking the Ice: Icebreaker Activities and Games

Building relationships from the outset is crucial for a cohesive and high-performing team. Icebreaker activities and games are excellent tools to break

down barriers, foster connections, and create a safe space for collaboration. Embrace these opportunities to bring out the best in your team. In my experience, I have used a simple breaker like "One fun fact about you" when I am connecting with team members for the 1st time. This activity helped to set the stage for our get-to-know-you conversation.

V.   Building Relationships with Team Members

Strong relationships are the cornerstone of successful teamwork. As a Scrum Master, building trust and rapport with your team members is vital. Take the time to get to know them outside on a personal level, listen actively, and show genuine interest in their well-being. Below is an example of how I have initiated a relationship-building opportunity in the past:

Scenario: Hello, team! As your friendly neighborhood Scrum Master, I believe in coffee breaks, not just Daily Scrum. Can you grab a coffee and chat about our passions, hobbies, and anything that isn't backlog-related?

As you embark on your Scrum Master's evolution in a new organization, remember to embrace each moment with confidence, authenticity, and a dash of humor. Introduce yourself with flair, navigate the organizational systems like a pro, break the ice with special activities, and build genuine relationships with your team members. Stay tuned for the next chapter, where we'll delve into preparing for success from day two onward. Cheers to an extraordinary evolution ahead!

CHAPTER 3
PREPARING FOR SUCCESS FROM DAY TWO AND MOVING FORWARD

In this chapter, we will embark on a thrilling evolution of discovery, learning, integration into the organizational system, and growth as we delve into the steps that will set you up for success from day two and beyond. Drawing from my extensive experience working with various organizations, I will share valuable insights and practical examples to guide you on this exciting adventure.

As we navigate this chapter, we will balance the delicate relationship between confidence and vulnerability while embracing our roles as servant leaders. Let us embark on this evolution with excitement, authenticity, simplicity, enjoyment, and fun.

I.   Learning about the Organization and Systems

As a Scrum Master joining a new organization, it is crucial to understand its culture, values, and systems. The organizational context in which you operate significantly impacts your ability to facilitate Agile practices effectively. Around day two, you would still be working on completing the onboarding and compliance training. I usually bookmark helpful resources that cover insights about the organization's culture, values, systems, and benefits. When I am more settled, less overwhelmed with so much information, and things begin to make sense, I usually revisit the saved resources with a clearer mind and read them for better understanding.

Scenario: When I joined a global medical device company, I was fascinated by its commitment to sustainability. To align Agile practices with their eco-conscious goals, I dedicated time to deep-dive into their core values, business acumen, overall vision, and goals from day two. By immersing myself in their sustainability initiatives and understanding their environmental impact aspirations, I could tailor Agile practices to resonate with their vision. This approach fostered increased engagement and a sense of purpose within the team.

II. Learning about Your Team Members, Key Stakeholders, and Building Relationships

Building solid relationships with your team members and key stakeholders is fundamental to your success as a Scrum Master. Invest time learning about their backgrounds, aspirations, and pain points to be sure that you are working with their agenda as your stakeholders and not your agenda as a Scrum Master. Helping your team and stakeholders improve their key pain points will foster collaboration, trust, and synergy.

Scenario: From day two, I leveraged the power of individual connection in one organization. I scheduled one-on-one meetings with each team member to understand their unique perspectives, work preferences, key pain points, and career goals. I gained profound insights into their motivations and challenges by actively listening and showing genuine interest. This personalized approach enabled me to tailor my support and coaching to meet their needs, fostering trust and open communication within the team.

Here are some questions I asked, categorized by accountabilities:

For Leadership (Managers, Business Stakeholders, etc.)

- What problem are you trying to solve?
- Why did you decide to adopt Business Agility?
- What are your expectations of me as your new Scrum Master?
- What is your stake in the team's work?
- What does success look like?
- What are your top 3 pain points?
- How will you measure my impact on the teams?
- What information or metrics do you find most valuable in creating alignment?
- Is there a structure in place for you to understand what value the team is delivering?
- Are there any particular communication preferences or channels we should consider when engaging with stakeholders?
- Are there currently any Agile Communities of Practices?

For Developers:

- What is your accountability on the team and area of strength?
- Give a high-level explanation of your software development process
- What is your most significant source of motivation?
- What are your expectations of me as your new Scrum Master?

- What are your key pain points?
- What would be your magic wand if I were to enable the team's improvement?
- How do you prefer to receive feedback as a team member?
- Are there any specific tools or platforms you find helpful for collaboration?
- Do you have any preferences for how technical discussions or code reviews are conducted?
- What is your perception of the Agile ways of working in this organization and your team?

For Product Owners:

- Which product are we working on?
- What is the product vision and goal?
- What does success look like?
- How do you prefer to communicate new product requirements or changes?
- What are your expectations of me as your new Scrum Master?
- Are there any methods or formats you find most effective for providing user stories or acceptance criteria?
- What level of involvement do you expect from the team during work item definition and elicitation?
- Which Agile framework does the team currently use?
- How would you like to be informed about sprint progress and any changes to the sprint plan?
- Are there any specific communication channels or tools you find most helpful in sharing updates or seeking clarification?
- How often does the team have retrospective meetings to discuss the sprint performance and align on improvement opportunities?
- Who are our customers?

By actively seeking inputs from team members, you will gain insights into their communication preferences and adapt your approach accordingly.

III.   Learning about the Product or Project Vision and Goal

Understanding what problems the product will solve and who the customers and users are is crucial for a Scrum Master. Learn about the intricacies, value proposition, customer need, and purpose for addressing the problem. These insights will empower you to guide the team effectively and contribute meaningfully to the team's success.

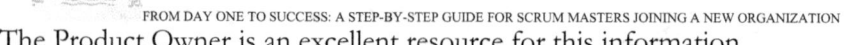

The Product Owner is an excellent resource for this information.

Scenario: I worked on an innovative solution to improve patient outcomes in a healthcare startup. To truly understand the vision behind the product, I engaged in discussions with Product Owners and Developers. By immersing myself in their world, I gained invaluable insights into the product's vision, purpose, the problems it aimed to solve, and the desired user experience. Armed with this in-depth knowledge, I guided the team in aligning their Agile practices with the product's strategic goals, resulting in an enhanced user experience and a deep sense of fulfillment within the team.

IV.   Understanding the Existing Agile Practices

Each organization has its unique Agile flavor, and understanding the existing Agile practices is vital for your role as a Scrum Master. By immersing yourself in the team's current practices, you can identify areas for improvement and leverage their strengths to drive successful Agile adoption.

Scenario: I joined a software development team three years into their Agile adoption. I observed their Scrum events, in which they only practiced Daily Scrum and sprint planning at the time. This observation helped me understand their collaboration patterns, bottlenecks, and improvement areas. I realized the team needed to become more familiar with Sprint Review, backlog Refinement, and Retrospective. Introducing a Sprint Retrospective event was one of the low-hanging fruit improvement action items I completed quickly because we needed an official platform to discuss and align on future improvement opportunities. By harnessing the team's collective intelligence and involving them in the process, we identified improvements to enhance their Agile events and empower them to reach new heights of their performance goals.

V.   Documenting Your Learning and Research Outcomes

As a Scrum Master, as you observe, listen, and ask questions, documenting your learning and research outcomes is essential for building a solid foundation and creating a knowledge repository that will guide your actions in the future. Capture key insights, trends, lessons learned, and best practices. This knowledge repository may become your Scrum Master backlog ordered in order of the team's priority pain points to ensure effective continuous learning and improvement.

Scenario: In my early days as a Scrum Master, I recognized the need for structured documentation. I created a centralized repository using Microsoft OneNote to capture my learnings, research outcomes, and practical tips I had gathered along the way. By organizing this information systematically, I had a wealth of knowledge at my fingertips, enabling me to refer back to valuable insights and share them with other Scrum Masters and teams. This approach saved time and fostered a culture of continuous learning and growth. It might be beneficial to store this type of information outside the company network (as long as it is not proprietary) so that you can leverage your learning if/when you leave.

VI.   Facilitating a Team Kickoff Workshop

Conducting an Agile team kickoff workshop as a newly hired Scrum Master is paramount in setting the stage for a successful journey with your new organization. This workshop is a crucial opportunity to unite the team, align their understanding of Agile principles, and establish a shared vision for collaboration and success.

The kickoff workshop provides a platform to introduce yourself as the Scrum Master and establish your role as a facilitator of Agile practices. It allows you to build rapport with team members, create a positive and inclusive atmosphere, and foster a sense of trust and psychological safety from the beginning. By actively listening to team members' perspectives and concerns,

25

you can address any apprehensions or misconceptions and create an environment where everyone feels heard and valued.

During the workshop, you can guide the team in defining or re-aligning their purpose, goals, and shared values. By facilitating discussions and exercises, you can encourage team members to articulate their expectations, identify areas of improvement, and collectively define success criteria for their Agile journey. This collaborative process fosters a sense of ownership and empowers team members to take responsibility for their work and the overall project outcomes.

Furthermore, the workshop allows the establishment of essential Agile events and practices. Through this interactive session, you can introduce concepts such as backlog refinement, sprint planning, Daily Scrum, sprint review, and retrospectives. This enables team members to start understanding how Agile events work, their significance in promoting transparency and continuous improvement, how these practices align with the organization's objectives, and set clear expectations.

Scenario: I vividly recall when I had the opportunity to conduct an Agile team kickoff workshop as a newly hired Scrum Master at a medical device Company. It was exciting for me to gather the team and set the stage for our Agile journey. Before the workshop, I reached out to each team member individually to introduce myself, express my eagerness to work with them and gather insights about their experiences and expectations.

On the workshop day, I arrived early to prepare the meeting room, ensuring a comfortable and welcoming environment. As the team members entered, I greeted them warmly, establishing a friendly and inclusive atmosphere right from the start. To break the ice and foster friendship, I kicked off the workshop with a fun activity where everyone shared an interesting fact about themselves.

The workshop agenda was carefully crafted to cover critical aspects of Agile principles, team dynamics, and project expectations. I facilitated interactive discussions and exercises to encourage active participation and collaboration. We collectively explored the values and principles that underpin Business Agility, discussing how they could be applied to our specific context.

I emphasized the importance of clear communication and collaboration during the workshop, introducing Agile events. I encouraged open dialogue, inviting team members to share their thoughts, concerns, and ideas. We

discussed ways to foster a positive team culture, promote accountability, and embrace continuous improvement. By the end of the workshop, the team had developed a shared understanding of Agile principles, established a foundation of trust and collaboration, and created a sense of excitement for the journey ahead.

The Agile team kickoff workshop proved to be a pivotal moment for our team. It set the stage for effective collaboration, alignment, and a shared commitment to Agile values. The workshop not only equipped the team with the necessary knowledge and tools but also fostered a sense of unity and enthusiasm. It was a truly empowering experience, and I continue to witness the positive impact of that workshop on our Agile journey as we work together toward our goals.

As we wrap up Chapter 3, you have gained valuable insights into the crucial steps to prepare for success as a Scrum Master from day two and beyond. Remember to embrace the balance of confidence and vulnerability, infusing authenticity and humor into your interactions. The evolution continues, and the next chapter will delve into the art of shadowing and observation. Join me in Chapter 4 as we explore the power of keen observation, listening, and asking questions to understand team dynamics and identify areas for improvement.

CHAPTER 4
SHADOWING AND OBSERVING FROM WEEK ONE

Welcome to Chapter 4. This exciting chapter will embark on a fascinating observation, exploration, and discovery journey. As a Scrum Master, mastering the science of observation is essential to gaining valuable insights into your team and organizational dynamics, communication styles, pain points, and areas for improvement. Drawing from my extensive experience of over 13 years working with different organizations, I will share real examples and authentic stories to guide you through this exciting chapter.

I.   Embracing the Role of an Observer

In your first weeks as a Scrum Master, embrace the role of an observer with an open mind and a genuine desire to understand. Step back and observe the team's interactions, dynamics, and processes. Observing allows you to gain a holistic view and identify opportunities for growth and improvement.

The Doctor Analogy: Imagine going to the hospital because you do not feel well. Typically, a doctor will not just diagnose you without gathering information about your symptoms and medical history, run some tests guided by data, then diagnose you based on the results. It is the same for Agile practitioners, which I call **'Agile Doctors.'** After joining a new organization, spending the first few weeks observing, actively listening, and asking open-ended questions is a great way to gather enough insights to guide you toward bringing the right solutions to your teams and your organization.

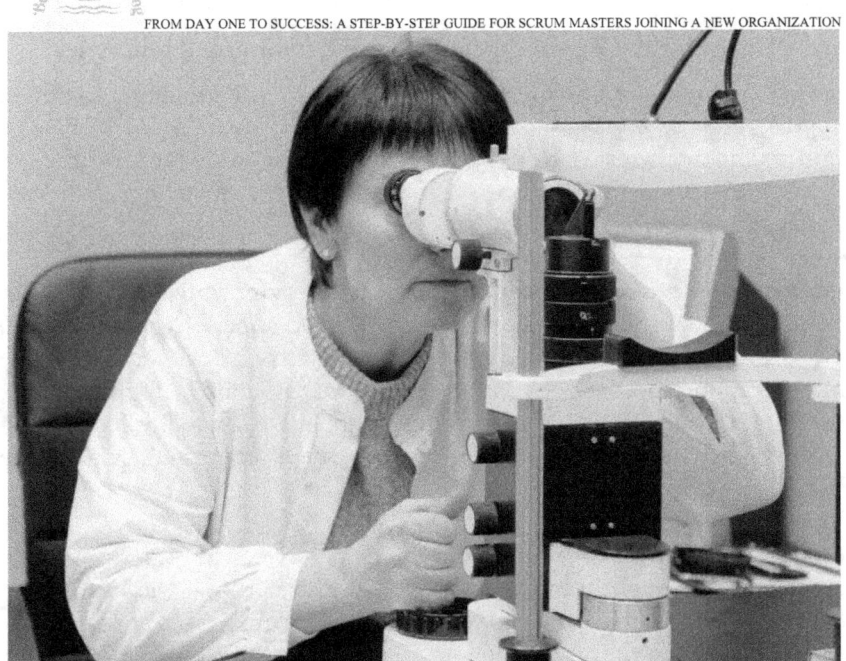

Scenario 01: As I joined a software development team, I enthusiastically adopted the Agile Doctor role. During their Daily Scrum, I closely observed how team members communicated, their body language, and the dynamics between individuals. By immersing myself in these observations, I discovered that some team members hesitated to speak up while others dominated the conversation. This keen awareness became a valuable foundation for future interventions and strategies to foster better collaboration and communication.

Scenario 02: In another organization, I had the opportunity to observe a team that I realized needed help with communication barriers. By consciously stepping back and observing their interactions during meetings, I noticed certain team members dominating the conversation while others remained quiet. This keen observation helped me identify an opportunity to foster balanced participation and create a safe space for everyone to contribute ideas.

II. Learning the Team Dynamics and Communication Styles

Teams are composed of unique individuals, each bringing their perspectives, strengths, and communication styles. Understanding these dynamics is critical to building trust, facilitating effective collaboration, and fostering a positive team culture right from the first few weeks of joining your new organization.

Scenario: I observed distinct communication styles among team members in a cross-functional team. Some preferred direct and concise communication, while others favored more elaborate explanations. I adjusted my communication approach by appreciating these individual preferences, ensuring everyone felt heard and understood. This adaptability fostered an environment where diverse voices were valued, contributing to more effective problem-solving and decision-making.

Scenario 02: I joined a team with a mix of introverted and extroverted personalities. Through keen observation, I discovered that the introverted team members often felt overshadowed during meetings, resulting in their ideas going unheard. By adapting my facilitation approach and creating opportunities for quieter voices to be heard, we were able to tap into a wealth of untapped ideas and perspectives.

So, it is imperative to meet the teams where they are. Through empathy and emotional intelligence, help bring them up to speed on their evolution to high performance.

III. Identifying the Current Pain Points and Areas for Improvement.

Every team has pain points and areas for improvement. As an astute observer, you identify these challenges and opportunities. This insight empowers you to address them proactively and positively impact team performance.

Scenario 01: In a highly dynamic project, I noticed that the team experienced frequent interruptions that affected their focus and outcome. By identifying this pain point, I initiated a discussion with the team to understand the root causes. Through open dialogue and collaborative problem-solving, we introduced strategies such as time-boxing meetings and establishing quiet work periods. These minor adjustments significantly reduced interruptions and enhanced the team's ability to deliver quality work.

Scenario 02: I observed a team needing more time management and usually achieved below 50% of their sprint goals. Through conversations with team members listening, and asking questions, I discovered that there needed to be more clarity in User Stories, and poor prioritization was contributing factor. Armed with this knowledge, I facilitated a retrospective session focused on improving their planning and prioritization practices. I also partnered with the Product Owner to coach him on product backlog prioritization techniques. By addressing these pain points, we increased the team's efficiency and delivered over 85% of the anticipated value within three months for the first time. This achievement made The team highly motivated, and the customers were impressed.

IV.   Capturing Observations and Insights

As you observe and gain valuable insights, capturing and documenting your observations is crucial. This process allows you to reflect on patterns, make connections, and refer to your findings when strategizing improvements.

Scenario: I habitually captured my observations and insights in a dedicated journal throughout my career. Whether it was a team member struggling with a particular task or a communication breakdown during a sprint review, I documented these experiences. This journal became a valuable resource, serving as a knowledge repository and enabling me to spot trends and patterns over time. I could detect improvement opportunities and achievements by reviewing my observations periodically. This information would be valuable in planning targeted interventions to address the gaps effectively. Some digital tools I have used include Microsoft OneNote, Google Drive, OneDrive, Confluence, etc.

As we wrap up Chapter 4, you have learned the importance of embracing the role of an observer, understanding team dynamics and communication styles, identifying pain points, and capturing valuable observations and insights. Remember to balance your confidence and vulnerability, infusing authenticity and humor into your leadership style. The evolution continues, and the next chapter will delve into building trust and credibility with your team members. Join me in Chapter 5 as we explore effective communication, active listening, and navigating difficult conversations with humor and tact.

CHAPTER 5
BUILDING TRUST AND CREDIBILITY

I.   Establishing Trust and Credibility with Team Members

Trust is earned through authenticity, transparency, and consistent actions. Establishing trust with your team members begins with building genuine relationships and demonstrating that you have their best interests at heart. Establishing credibility is different from establishing trust. Credibility comes with walking the talk and staying consistent. Once you earn that trust, you must back it up with action, helping them resolve some of their key pain points. Once your teams and organizations see that you are actionable and valuable, your credibility will be earned automatically.

Scenario 01: Let me share a personal experience: In one organization I worked with, a team was hesitant to trust a new Scrum Master due to past negative experiences. I approached the situation humbly, actively listening to their concerns and acknowledging their experiences. I slowly gained their trust by showing genuine empathy and committing to being their advocate. Through open and honest communication, we built a strong foundation of trust that paved the way for successful collaboration. From week 3 of joining that organization, I also helped to repurpose the team's Daily Scrum and adjusted the meeting time from one hour to 15 minutes. The team was impressed as the Daily Scrum became highly engaged and efficient.

Scenario 02: I once worked with a team where trust had been eroded due to a lack of transparency in decision-making. To rebuild trust, I initiated regular one-on-one conversations with team members, actively listening to their concerns and seeking their input. By openly addressing their questions, addressing any misunderstandings, and empowering them to be part of the decision-making process, I gradually rebuilt trust and reestablished a strong foundation for collaboration.

II.   Communicating Effectively and Actively Listening

Effective communication is vital in establishing trust. As a Scrum Master, communicating concisely and empathetically is essential. However, active listening is equally crucial—being fully present and engaged when team members share their thoughts, concerns, and ideas. Team members can tell when they're conversing with you as a Scrum Master, and you are not listening from your heart.

Scenario 01: I started a counseling program, and during my first one-hour session with the counselor, he yawned the whole time and continuously rushed me to wrap up when I was in the process of communicating my challenges. Another thing that I noticed was that he completed most of my sentences even without knowing what I was about to say. I left that session feeling like I wasted my time, energy, and emotions.

Scenario 02: During a retrospective, a team member expressed frustration with the need for more clarity in project goals. Instead of brushing off the comment, I actively listened, seeking to understand the underlying issues. By addressing the concerns and ensuring transparent communication about project objectives, the team member was heard and valued by me. This not only strengthened our working relationship but also improved overall team morale.

Scenario 03: During a sprint retrospective, tensions were high due to conflicting viewpoints within the team. I encouraged open dialogue instead of rushing to resolve the conflict, ensuring everyone could express their perspectives. By actively listening to each team member's concerns and facilitating a respectful exchange of ideas, we reached a consensus that addressed everyone's needs and ultimately strengthened the team's bond.

III.   Balancing Transparency and Confidentiality

As a Scrum Master, you may often be privy to sensitive information. It is crucial to navigate the delicate balance between transparency and confidentiality. Strive to be transparent about the information that can benefit the team while respecting the confidentiality of personal matters or sensitive discussions.

Scenario 01: In one instance, a team member privately confided in me about personal challenges impacting their performance. While it was essential to maintain confidentiality, I also realized that transparent communication with the team about the member's situation could foster empathy and support. We balanced respecting privacy and promoting a supportive team environment
 through open dialogue and careful handling.

Scenario 02: I was entrusted with confidential information regarding upcoming organizational changes in one organization. However, I knew these changes would affect the team as some team members were to be let go. I organized a special meeting with the team, where I emphasized the importance of confidentiality and explained that I had information that I could not disclose at that time. By acknowledging their curiosity and reassuring them that I would share relevant information as soon as possible, I maintained trust and transparency while respecting the boundaries of confidentiality.

IV.   Navigating Difficult Conversations with Humor and Tact

Difficult conversations are an inevitable part of being a Scrum Master. Navigating these conversations with humor and tact can help ease tension and foster a positive atmosphere. Using humor appropriately can create a safe space where team members feel comfortable addressing challenging topics. Naturally, people find me humorous, and I get fulfillment in being the light in the room. I have leveraged my humorous super-power throughout my career as a Scrum Master to make work enjoyable for my team members.

Scenario 01: In a retrospective, tensions arose as team members expressed frustration with a particular process. Instead of letting the tension escalate, I injected humor into the conversation. By lightening the mood, we could address the issues constructively and find solutions collectively. Humor helped defuse tension and allowed everyone to approach the discussion with an open mind.

Scenario 02: I once found myself in a situation where a team member consistently missed sprint commitments, causing frustration within the team. Instead of confronting, I approached the conversation with humor and empathy. I invited the team member for a one-on-one discussion, using a lighthearted analogy to highlight the impact of their missed commitments. This approach allowed us to address the issue constructively without placing blame or creating defensiveness. Together, we identified strategies to improve commitment reliability, and the team member felt supported rather than criticized.

As we conclude Chapter 5, you have gained valuable insights into building trust and credibility as a Scrum Master. Establishing trust and credibility with team members, communicating effectively, balancing transparency and confidentiality, and navigating difficult conversations with humor and tact will strengthen your relationships and create an environment conducive to growth and success. Remember, being a servant leader requires both confidence and vulnerability, and by embodying these qualities, you will inspire trust and credibility and guide your team toward remarkable achievements.

In the next chapter, we will explore the art of creating, measuring, and tracking performance goals, empowering you to drive continuous improvement and growth within your team.

CHAPTER 6
GUIDING AGILE ADOPTION

As a Scrum Master, guiding Agile adoption by helping the team embrace Agile principles and practices is a critical part of your job around the first 30 to 90 days of joining the organization. An excellent place to start is by assessing the team's Agile maturity level and identifying opportunities for improvement. This chapter will explore four key aspects: assessing Agile maturity, identifying incremental improvements, implementing best practices, and fostering continuous learning and growth.

I.   Assessing the Team's Agile Maturity Level:

Before guiding the team towards Agile adoption, assessing their readiness and receptiveness to change is essential. Evaluate their understanding of Agile principles, current practices, and willingness to embrace Agile values.

I often start by assessing their understanding and application of Agile principles, practices, and values. This assessment helps identify areas where the team excels and areas that require further development. Based on the outcome of the evaluation, I then partnered with the team to prioritize the gap and start taking action toward improvement.

Scenario 01: I used a maturity assessment questionnaire that covers various aspects of Agile, such as roles and responsibilities, collaboration, transparency, and feedback loops. Analyze the results to understand the team's strengths and areas for improvement.

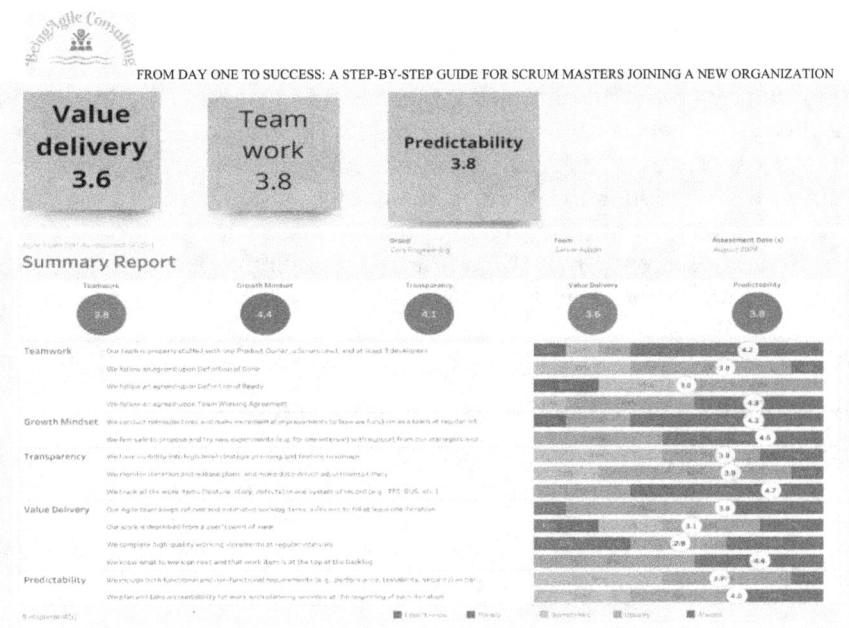

Scenario 02: In another organization, I conduct interviews or surveys to gauge the team's knowledge of Agile, their level of collaboration, and their openness to change. This assessment helped me understand the team's starting point and identify areas where additional support may be required.

II.   Identifying Incremental Opportunities for Improvement:

Once you have assessed the team's Agile maturity, identify incremental opportunities for improvement. Look for specific practices or processes that can be enhanced to increase effectiveness, efficiency, and value delivery.

If the assessment reveals a need for more clarity in user stories, work with the Product Owner and developers to implement techniques such as user story mapping or refining the Definition of Ready. These incremental improvements contribute to a smoother workflow and better outcomes.

III.   Introducing Agile Best Practices with Practical Examples:

Introducing Agile best practices is vital for success. Select relevant practices that align with the team's needs and provide practical examples to illustrate their application. Demonstrate how these practices can address common challenges and enhance the team's performance.

Scenario: In the past, I realized the team was not leveraging the Scrum events, which caused misalignment and a lack of transparency on the team.

The sprint outcome and team morale was negatively impacted. I introduced Backlog Refinement sessions, Daily Scrums, Sprint planning, Sprint Reviews, and Retrospectives to facilitate collaboration, improve transparency, and encourage continuous improvement. Share real-world examples of how these practices have positively impacted teams. We quickly synchronized and increased transparency by 60% within three months, leading to a 45% improvement in value delivery.

IV.   Encouraging Continuous Learning and Growth:

Enable an environment that encourages continuous learning and growth. Foster a culture where team members feel empowered to experiment, learn from failures, and acquire new skills. Provide opportunities for team retrospectives, training, workshops, and knowledge sharing. The image below is an extract from a BeingAgile Consulting team retrospective:

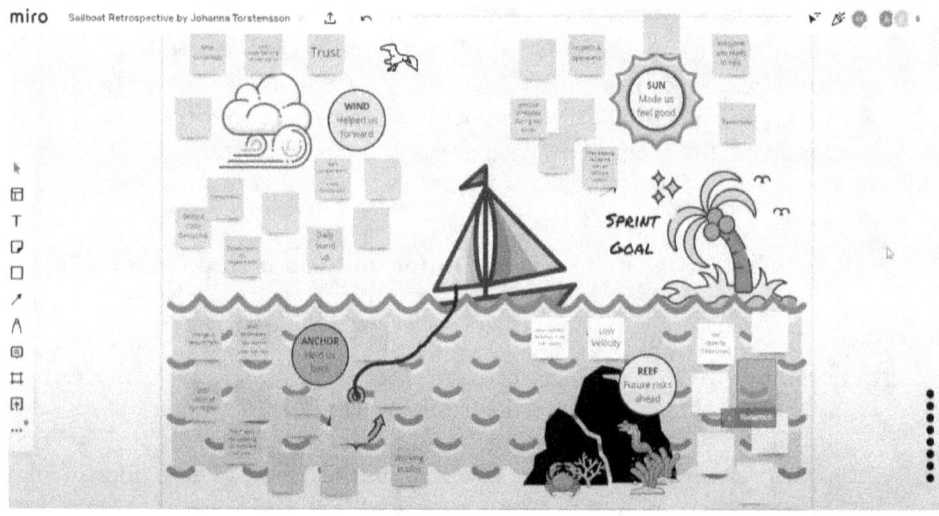

Scenario: Organize lunch and learn sessions, where team members can share their expertise or explore new topics related to Agile. Encourage individuals to attend conferences and webinars or join professional communities to stay updated on industry trends and best practices.

By assessing Agile maturity, identifying incremental improvements, implementing best practices, and fostering continuous learning and growth, you can drive the team towards higher performance and sustained success as a Scrum Master. These efforts create an environment of continuous improvement, collaboration, and innovation that supports the team's evolution in embracing the Agile ways of working.

CHAPTER 7
CREATING, MEASURING, AND TRACKING PERFORMANCE GOALS

Setting and tracking performance goals is crucial for Scrum Masters to continuously improve their effectiveness and contribute to the team's and the organization's success. This chapter will explore four key aspects of goal setting and tracking: establishing KPIs, creating action plans, utilizing metrics and data, and celebrating achievements.

I.   Establishing KPIs for Yourself and When to Set Them:

Setting Key Performance Indicators (KPIs) is a proactive approach to defining the desired outcomes and measuring your progress as a Scrum Master. When setting KPIs, consider your accountabilities, responsibilities, and the organization's objectives. It is recommended to set KPIs at the start of a new project or after assessing the team's current state, identifying opportunities for improvement. Set your performance goals around specific areas for improvement.

Example: As a Scrum Master, you may set a KPI to:
- Enhance team efficiency rating by 10% within the next quarter." This KPI focuses on improving the team's processes and can be established during the team's retrospective meeting at the beginning of a project.
- Improve team collaboration and reduce the time taken to deliver product increments. I set specific KPIs, such as increasing team engagement in retrospective activities by 20% within 12 weeks.
- Reduce the average sprint cycle time by 20%.
- Increase the team's average sprint predictability by 20% within six sprints.

By defining these objectives and KPIs, I aligned my efforts with the organization's goals and tracked my impact over time. I leveraged them to ensure I enabled a positive and value-adding change within the team.

II.   Creating Action Plans and Tracking Progress:

Creating action plans lets you outline the steps and initiatives required to achieve your goals. Break down larger objectives into smaller, manageable tasks and assign time boxes to track progress effectively. Regularly review and update your action plans to adapt to evolving circumstances.

Scenario: If your goal is to enhance communication within the team, your action plan may include introducing a Daily Scrum, conducting workshops on effective communication, and encouraging cross-functional collaboration. Track your progress by monitoring the frequency and quality of team interactions.

III.   Utilizing Metrics and Data to Measure Success:

Metrics and data provide objective insights into your performance as a Scrum Master. They help you assess your actions' impact and identify improvement areas. Select relevant metrics aligned with your goals and regularly collect data to measure progress.

Scenario: Use metrics such as team velocity, sprint burndown charts, customer satisfaction ratings, or lead time to assess the effectiveness of your Agile practices and identify bottlenecks or areas requiring further attention. Analyze these metrics to gain valuable insights and adjust your approach accordingly.

By defining these objectives and KPIs, I aligned my efforts with the organization's goals and tracked my impact over time. This approach has earned me some cool promotions and recognition. 😊

IV.   Celebrating Achievements and Adjusting Goals as Needed:

Celebrating achievements is an essential component of performance tracking. Recognizing little wins not only boosts team morale but also motivates continuous improvement. Additionally, be open to adjusting your goals based on feedback, changing circumstances, or new priorities.

Scenario: When the team successfully delivers a high-quality product with minimal defects, celebrate their achievement and acknowledge their hard work. Adjust your goals based on lessons learned and emerging requirements to ensure they remain relevant and contribute to the team's success.

CHAPTER 8
FACILITATING ENGAGING MEETINGS AND FOSTERING A POSITIVE TEAM CULTURE

As a new Scrum Master, usually around the second month of joining your organization, you should be more comfortable and nicely integrated into your team(s). One of the critical aspects of your accountabilities is facilitating engaging meetings and fostering a positive team culture. This chapter will share my real-world experiences and practical examples of effectively accomplishing these goals.

I.   **Planning and Facilitating Engaging Backlog Refinement sessions, Sprint Planning, Daily Scrums, Retrospectives, and Sprint Reviews events:**

Engaging the team during these meetings requires careful planning and a focus on active participation. First, I ensured that the purpose of the scrum events was well understood and that every team member was aligned with the process. For example, during backlog refinement sessions, I encourage the team to collectively refine user stories by leveraging their expertise and insights. Involving everyone in the discussion ensures the backlog items are well understood and adequately estimated.

During sprint planning, I guide the team in breaking down user stories into actionable tasks as needed, encouraging collaboration and shared responsibility. This helps create a clear plan for the sprint and promotes a sense of ownership among the team members.

In Daily Scrums, I emphasize the importance of concise and focused discussions. I encourage team members to share their progress, highlight obstacles, and collaborate on problem-solving during the parking lot to keep the energy high. I empowered them to own this meeting and take turns in facilitating. I shielded the team from going off the meeting agenda and ensured every developer and other team member understood Who-Does-What during this meeting. Maintaining a time-boxed and engaging Daily Scrums ensures the team remains aligned and motivated.

During retrospectives and sprint reviews, I facilitate open and honest discussions. I create a safe space for the team to reflect on their performance, share their observations, and suggest improvements. We foster a continuous improvement and learning culture by celebrating achievements and acknowledging challenges.

II. Using Facilitation Techniques to Drive Collaboration and Great Outcomes:

Facilitation techniques are crucial in enabling collaboration and productivity during meetings. Employing these techniques enhances engagement and helps the team make better decisions.

For instance, I use visual collaboration tools such as virtual whiteboards or online sticky notes during meetings. These tools allow the team to actively participate and contribute their ideas, ensuring everyone's voice is heard and valued.

Brainstorming sessions have been incredibly effective in encouraging innovative thinking. The team can explore multiple possibilities and develop creative solutions by providing a structured platform for generating ideas. I often use mind mapping or storyboarding techniques to visualize and organize these ideas, fostering a collaborative environment.

Techniques like dot voting or a fist-of-five are highly effective in decision-making. These methods enable quick prioritization and consensus-building, ensuring the team's collective wisdom drives decision-making.

III. Injecting Humor and Creativity into Meetings:

Humor and creativity can transform boring meetings into engaging and enjoyable experiences. Incorporating these elements helps create a positive atmosphere and encourages active participation from the team.

I often start meetings like Retrospective events with a fun icebreaker activity or share relevant and lighthearted anecdotes to inject humor. Setting a positive tone makes team members feel more relaxed and open to expressing their ideas and opinions.

Creativity can be nurtured by introducing interactive exercises during meetings. For example, I used mind-mapping techniques during a problem-solving workshop to encourage out-of-the-box thinking and inspire innovative solutions. Role-playing or storyboarding activities also stimulate creativity and provide a fresh perspective on problem-solving.

IV. Managing Conflict and Fostering a Positive Team Culture:

Conflict is inevitable within any team, but managing it effectively is essential for maintaining a healthy team culture. Over the years, I have learned valuable lessons in addressing conflicts and creating an environment of trust and collaboration.

When conflicts arise, I encourage team members to express their concerns and perspectives openly and respectfully. As a facilitator, I apply active listening and empathy, ensuring each team member feels heard and understood. Throughout the conflict resolution session, I stayed neutral and objective and encouraged the parties to own the process and the outcome. This approach promotes healthy discussions and helps resolve conflicts constructively.

Establishing team norms or working agreements is another powerful way to foster a
positive team culture. By collaboratively defining expected behaviors and values, we create a shared understanding and a framework for respectful interactions. Regularly revisiting these norms and addressing deviations ensures the team maintains a positive and supportive environment.

In conclusion, by planning and conducting engaging meetings, using facilitation techniques, injecting humor and creativity, and managing conflict effectively, Scrum Masters can foster a positive team culture and drive excellent outcomes. These practices have been instrumental in my evolution as a Scrum Master, and I encourage you to adapt and apply them to your unique organizational context.

CHAPTER 9
CONFLICT MANAGEMENT AND TEAM COLLABORATION

In the evolution of being a Scrum Master, managing conflict and fostering team collaboration are vital skills. Reflecting on my experiences, I recall specific instances where these skills played a crucial role in enhancing team dynamics and enabling success right around the eleventh week of joining my new organization. Let me share examples from my real-world experiences on how I effectively managed conflicts, addressed team dysfunctions, encouraged open communication and collaboration, and built a positive team culture within my teams.

I.   Recognizing and Addressing Team Dysfunctions

Recognizing and addressing team dysfunctions is crucial for Scrum Masters when joining a new organization. As the facilitators of the Scrum framework, Scrum Masters play a pivotal role in ensuring effective teamwork and maximizing the team's potential. Team dysfunctions can manifest in various ways, such as lack of trust, poor communication, resistance to change, or conflicting priorities. Scrum Masters need to identify these dysfunctions early on, observe team dynamics, and foster an environment of open dialogue and trust. By encouraging transparency and providing a safe space for team members to express their concerns, Scrum Masters can address these dysfunctions proactively. They should facilitate team-building activities, establish clear communication channels, and promote collaboration among team members. Additionally, Scrum Masters can help the team understand the benefits of agile principles and guide them through any necessary adjustments or changes. By promptly and effectively addressing team dysfunctions, Scrum Masters can empower their teams to achieve high performance, drive innovation, and deliver exceptional results in their new organization.

Scenario: In my previous company, our team encountered a persistent issue where decision-making processes could have been faster and more cohesive. This dysfunction led to inefficiencies and a need for more alignment among team members. Recognizing the negative impact of this dysfunction on our sprint outcomes, I initiated a problem-solving workshop focused on understanding the root causes and finding solutions using the Cause and Effect problem-solving technique.

During the workshop session, I facilitated a discussion where team members openly shared their observations and concerns. We identified communication gaps, overlapping roles, and a need for more clarity in decision-making authority as critical dysfunctions. To address these issues, we collectively brainstormed and implemented practical solutions. We revised our communication channels, clarified roles and responsibilities, and introduced clear decision-making frameworks. By recognizing and addressing these dysfunctions, our team became more aligned, decision-making improved, and collaboration flourished.

II. Facilitating Conflicts with Empathy and Respect

Conflicts are inevitable in any team dynamic. As a Scrum Master, I embraced the role of a facilitator to guide the parties involved toward a creative and constructive resolution process which helped to maintain a harmonious work environment. In one instance, two team members had differing opinions on the implementation approach for a critical feature, leading to a heated argument during a sprint planning session.

To address the conflict, I stepped in as a neutral facilitator, enabling a safe space for both individuals to express their perspectives. I actively listened to each person, ensuring they felt heard and understood. With empathy and respect, I encouraged open dialogue and helped them uncover shared goals and common ground.

Through guided discussion and compromise, they reached a solution that integrated elements from both approaches. By acknowledging their efforts and highlighting their shared commitment to the project's success, I helped them build rapport and strengthen their working relationship. This conflict resolution resolved the immediate issue and created an environment where team members felt comfortable addressing conflicts openly, fostering a culture of trust, collaboration, and creativity.

III.   Encouraging Open Communication and Collaboration

Open communication and collaboration are essential for effective teamwork. As Scrum Masters facilitate the Scrum framework, they are responsible for fostering an environment where effective communication and collaboration thrive. Open communication involves promoting transparency, active listening, and encouraging team members to express their ideas, concerns, and challenges openly. By creating a safe space for open dialogue, Scrum Masters can help build trust among team members, allowing them to share their thoughts and opinions freely. Collaboration, on the other hand, entails promoting teamwork, cooperation, and shared responsibility. Scrum Masters should encourage cross-functional collaboration, where individuals with diverse skills and perspectives work together towards a common goal. They can facilitate collaborative activities such as daily scrums, sprint planning, and retrospectives to ensure everyone's input is valued and incorporated into the team's decision-making process. By emphasizing open communication and collaboration, Scrum Masters empower their teams to leverage team members' collective intelligence and creativity, leading to improved problem-solving, higher-quality increments, and increased team satisfaction. Ultimately, by fostering a culture of open communication and collaboration in the new organization, Scrum Masters enable the team to achieve higher performance, innovation, and success.

Scenario 01: In a Kanban project where our team faced challenges in knowledge sharing and collaboration, I took proactive steps to encourage openness and collaboration among team members. To promote open communication, I introduced Daily Scrum meetings as a platform for team members to share progress and challenges and seek assistance. I facilitated discussions where individuals could express their ideas and concerns freely. This was a Kanban team, but we realized the need to borrow some Scrum concepts like openness, courage, focus, respect, and commitment. Encouraging active listening and respectful feedback ensured that every team member felt valued and heard.

Additionally, I introduced collaboration tools and techniques such as pair programming and peer code reviews to foster knowledge sharing and collective ownership of the codebase. These practices encouraged cross-functional collaboration and enabled team members to leverage each other's expertise, leading to improved code quality and faster problem-solving.

IV.   Building a Positive Team Culture

Creating a positive team culture is essential for maintaining team motivation, engagement, and overall success.

Scenario 01: I observed low team morale in one project, resulting in reduced sprint outcomes and enthusiasm. When I joined the organization, I observed that team members had no other activities outside of work. Team members had no personal relationship with each other, and it seemed like they walked on eggshells.

To address this, I focused on building a positive team culture. I initiated team-building activities such as team lunches, offsite outings, and gamified retrospectives to foster camaraderie and a sense of belonging. These activities allowed team members to connect on a personal level, strengthening their bonds and building trust.

Furthermore, I encouraged celebrating individual and team achievements, big and small, to boost morale and create a positive work environment. Recognizing and appreciating team members' efforts and accomplishments increased their motivation and inspired others to strive for excellence.

After incorporating these practices, I witnessed a positive shift in the team's dynamics. Collaboration improved, healthy conflicts were embraced, and team members actively supported and uplifted one another. The positive team culture contributed to increased productivity, higher job satisfaction, and project success.

Scenario 02: In a geographically distributed team, I recognized the importance of fostering a positive team culture despite the physical distance. To create a sense of unity and team spirit, I organized virtual team-building activities and social events. For instance, I initiated a weekly "Coffee Chat" session where team members from different locations would come together for informal conversations, sharing personal experiences and discussing non-work-related topics. This created a relaxed and friendly environment, allowing team members to connect on a personal level and build relationships beyond work. Additionally, I encouraged using collaboration tools like Slack or Microsoft Teams for non-work-related channels, such as a "Watercooler" channel, where team members could share interesting articles and funny memes or engage in casual conversations. These initiatives helped bridge the geographical gap, fostered a positive team culture, and strengthened the sense of belonging among team members.

Scenario 03: In another geographically distributed team, I proactively focused on building trust and encouraging open communication among team members. To create a positive team culture, I introduced a practice called "Recognition Fridays." Every Friday, team members were encouraged to publicly recognize and appreciate their colleagues' efforts, achievements, or acts of support in a dedicated channel or during team meetings. This simple but powerful gesture helped foster a culture of appreciation, gratitude, and recognition. It boosted team morale and promoted a supportive and collaborative atmosphere.

Additionally, I organized virtual team retrospectives where team members could openly share their feedback, ideas, and concerns. Using online collaboration tools with features like anonymous feedback, I ensured everyone had a voice and could express their thoughts without fear. This created a safe space for open dialogue, problem-solving, and continuous improvement, ultimately contributing to a positive team culture even in a geographically distributed setting.

By recognizing and addressing team dysfunctions, mediating conflicts with empathy and respect, encouraging open communication and collaboration, and building a positive team culture, I have experienced firsthand the transformative impact these practices can have on team dynamics and overall performance.

CHAPTER 10
OVERCOMING CHALLENGES AND PITFALLS

As a Scrum Master, you will face numerous challenges and pitfalls in your evolution. These may become valuable lessons learned based on your attitude toward navigating these challenges. In this chapter, I will share my real-world experiences dealing with resistance and pushback, addressing team dysfunctions with authenticity and humor, navigating organizational politics and bureaucracy, and maintaining resilience and self-care.

I.   Dealing with Resistance and Pushback

Dealing with resistance and pushback is a significant challenge that Scrum Masters may face when joining a new organization. As you facilitate the adoption of Business Agility and the Scrum framework, they may encounter individuals or teams who are negatively reacting to change or skeptical about the benefits of Business Agility.

It is important to be patient, persistent, and adaptable. Change takes time, and it may require ongoing communication, coaching, and mentoring to help individuals and teams embrace new ways of working. You can gradually overcome resistance and pushback by providing continuous support, guidance, and feedback, fostering a culture of collaboration and continuous improvement.

Scenario: When introducing Agile practices to a new team, I encountered resistance and skepticism from some team members who needed to be more open to adapting their ways of working.

I took a proactive approach to address this challenge by seeking to understand their concerns and perspectives. I organized individual meetings with team members who expressed resisting behaviors to listen to their apprehensions and provide them with a safe space to express their thoughts. I built trust and established open communication by empathizing with their concerns and demonstrating genuine interest.

To overcome this pushback, I highlighted the benefits and positive outcomes of the specific Agile practices I introduced then. I shared success stories from other teams and conducted training sessions to educate the team on the concept. I suggested an experimentation approach to try out these changes incrementally. I gradually won over the skeptics and gained their support by providing clear explanations and addressing their concerns.

II.   Addressing Team Dysfunctions with Authenticity and Humor

Teams are not immune to dysfunctions, and addressing them promptly is vital for maintaining a healthy team dynamic. Incorporating authenticity and humor can be effective in addressing and resolving team dysfunctions.

Scenario: In one project, our team faced frequent breakdowns in communication, leading to misunderstandings and delays. I organized a theme-driven retrospective focusing on team communication to tackle this issue. During the retrospective, I introduced a fun and interactive activity where team members had to communicate using exaggerated gestures and facial expressions. 😄

This lighthearted approach broke the ice and brought a sense of humor into the discussion. It allowed team members to reflect on their communication styles and identify areas for improvement. By addressing the issue with authenticity and humor, we created an environment where team members felt comfortable acknowledging their shortcomings and working together to enhance communication.

III.   Navigating Organizational Politics and Bureaucracy

Organizational politics and bureaucracy can be challenging but inevitable. As a Scrum Master, you must understand the organization's culture and find ways to work within it. To be successful as a Scrum Master, you need to be seen not just within your team but outside of your team as well. Here are a few practical tips on navigating organizational politics and bureaucracy:

Build Relationships and Seek Allies: One way to navigate organizational politics is to build positive relationships and seek allies within the organization. Take the time to connect with key stakeholders, decision-makers, and influential individuals who can support the adoption of Business Agility. Engage in informal conversations, understand their concerns, and align your goals with theirs. By fostering strong relationships, you can gain support, navigate political obstacles, and create a network of allies who can advocate for the Agile ways of working. These allies can help you navigate bureaucratic processes, streamline decision-making, and overcome resistance from within the organization.

Educate and Influence: Bureaucratic processes can often hinder the agility and flexibility required for Scrum. As a Scrum Master, it is vital to educate others about the benefits of the Agile ways of working and how they can positively impact the organization's goals. Take the initiative to organize workshops, lunch and learns, or presentations to share success stories, case studies, and empirical evidence of Agile's effectiveness. You can influence stakeholders and decision-makers to consider adopting agile practices by

showcasing the tangible benefits. Additionally, be prepared to address regulations, compliance, or other bureaucratic requirements concerns. Find ways to work within the existing framework while advocating for agility, demonstrating how they can coexist and enhance the organization's processes and outcomes.

Be adaptable: Be willing to adapt your approach to the organization's culture and practices. This will help to make your initiatives more palatable to the organization and increase their chances of success. This is why it is essential first to understand the organization's priorities, then build a partnership with stakeholders and work by their agenda, not your agenda, as a Scrum Master.

Scenario: I proactively sought opportunities to connect with key organizational stakeholders and decision-makers. By understanding their goals, concerns, and motivations, I position Agile practices as a solution that aligns with the organization's objectives.

In addition, I collaborated with other Scrum Masters and Agile champions across the organization to share best practices and learn from their experiences. By building a network of support and leveraging collective knowledge, I gained valuable insights into navigating organizational politics and bureaucracy.

Remember, navigating organizational politics and bureaucracy requires a diplomatic and patient approach. Understanding the organizational context, identifying key influencers, and tailoring your communication and strategies are important.

IV.   Maintaining Resilience and Self-Care as a Scrum Master

Being a Scrum Master can sometimes be demanding and emotionally draining, so resilience and self-care are crucial for Scrum Masters. Here are two examples of prioritizing resilience and self-care:

Establish Boundaries and Manage Workload: As a Scrum Master, it is important to establish clear boundaries and manage your workload effectively. Understand your capacity and set realistic expectations with stakeholders and team members. Avoid taking on too many responsibilities or overcommitting yourself, as it can lead to burnout and reduced effectiveness. Learn to empower team members, ask for help as needed and encourage self-organization within the team. By managing your workload and setting boundaries, you can maintain a healthier work-life balance, reducing the risk of exhaustion and stress.

Seek Support and Continuous Learning: Navigating diverse human behaviors can be challenging, and seeking support and continuously learning to maintain resilience is essential. Connect with other Scrum Masters or Agile Practitioners within or outside the organization to share experiences, seek advice, and gain insights. Join agile communities or attend meetups, conferences, or webinars to stay updated on industry trends and best practices. Engage in regular introspections or self-reflection to identify areas for improvement and adjust your approach accordingly. By seeking support and continuously learning, you can enhance your skills, stay motivated, and overcome challenges more effectively.

Additionally, prioritize self-care activities outside of work. Engage in hobbies, exercise regularly, get enough sleep, and maintain a healthy diet. Take breaks during the workday to recharge and disconnect from work-related tasks. Remember to practice mindfulness or meditation techniques to reduce stress and promote mental well-being. By prioritizing self-care, you can enhance your overall resilience, focus, and ability to handle the demands of your role as a Scrum Master.

Scenario: I have prioritized self-care strategies to strengthen my resilience and ensure my well-being. I established clear boundaries by setting aside time for personal activities and hobbies outside of work. Whether practicing mindfulness, engaging in physical exercise, or spending quality time with loved ones, these activities helped me recharge and maintain a healthy work-life balance.

Additionally, I fostered a supportive network by connecting with fellow Scrum Masters and joining Agile communities. These connections provided opportunities for knowledge sharing, mentorship, and emotional support.
By prioritizing self-care, I brought my best self to the team and effectively supported them in their Agile evolution.
Remember, maintaining resilience and self-care is not a one-time effort but an ongoing practice.

In conclusion, overcoming challenges and pitfalls as a Scrum Master requires resilience, adaptability, and a willingness to navigate various obstacles. I have

grown professionally and personally by dealing with resistance and pushback, addressing team dysfunctions with authenticity and humor, navigating organizational politics and bureaucracy, and maintaining resilience and self-care. I hope my experiences and insights in this chapter will empower you to tackle challenges head-on and thrive as a Scrum Master.

DO NOT AVOID CHALLENGES 😱! Embrace them and enjoy the learning and improvement evolution.

CHAPTER 11
CONTINUOUS IMPROVEMENT AND GROWTH

As a Scrum Master, fostering a culture of continuous improvement and growth within your team and yourself is vital. This chapter will explore four key aspects contributing to my ongoing success as a Scrum Master.

I.   Celebrating Big and Small Wins

Celebrating big and small wins is a great practice to boost team morale and motivation. It provides a sense of accomplishment and encourages a positive mindset.

Scenario: In one of my previous organizations, we had been working tirelessly on a complex feature for several sprints. I organized a small celebration as the team delivered the whole increment successfully. I arranged a team lunch where we shared our achievements, acknowledged each member's contributions, and celebrated our collective success. It created a sense of unity, belonging and inspired the team to continue working towards our goals.

II. Encouraging a Culture of Innovation and Experimentation

Teams need to embrace innovation and experimentation to stay ahead in today's fast-paced world. As a Scrum Master, an aspect of your accountability is fostering this team culture.

Scenario: In one of my teams, we set aside dedicated time during each sprint for 'Innovation Fridays.' These days, team members can explore new ideas, technologies, or process improvements aligned with our project goals. We encouraged them to share their learnings and insights with the rest of the team. This sparked creativity and a sense of ownership, resulting in valuable innovations that enhanced our product. This practice also helped my team become cross-functional, and every team member knew a little of every skill, enough to back up each other and deliver quality results.

III. Seeking Feedback and Reflecting on Personal Growth

Continuous improvement applies to the team and your personal growth as a Scrum Master. Seeking feedback and reflecting on your performance is essential for professional development. Here's an example of how I embraced feedback and reflection:

Scenario: I supported a team of 5 in one of my previous organizations. Around my ninetieth day working with the team, I conducted individual feedback sessions with team members to understand their experiences working with me. I asked specific questions about my facilitation skills, communication style, and ways to support them better. I conducted this feedback solicitation exercise every three months. Additionally, I set aside time for personal reflection, keeping a journal to record my learnings, challenges, and areas for improvement. This allowed me to identify patterns, adjust my approach, and evolve as a Scrum Master."

IV. Evolving as a Scrum Master: From Day One to Ongoing Success

Becoming an effective Scrum Master is an ongoing evolution. It involves continuously learning, adapting, and evolving.

Scenario: I actively sought opportunities to expand my knowledge and skills to ensure my professional growth. I attended Scrum conferences, participated in workshops and webinars, and joined online communities to connect with other Scrum Masters. Additionally, I sought mentorship from

experienced Agile practitioners who provided guidance and insights to help me navigate challenges. By constantly seeking new perspectives and staying up to date with industry trends, I was able to bring valuable insights and fresh ideas to my team."

Celebrating wins, encouraging innovation, seeking feedback, and embracing personal growth, helped me to continuously evolve as a Scrum Master and guide my team towards ongoing success.

CHAPTER 12
CELEBRATING SUCCESS AND LEAVING A POSITIVE LEGACY

As your evolution as a Scrum Master continues in the organization, it's important to recognize achievements, leave a lasting impact, transition responsibilities, and extend well wishes for ongoing success. In this final chapter, we will explore these four key aspects that will help you celebrate success, leave a positive legacy, and ensure a seamless transition for the future.

I.   Recognizing Team Achievements and Little Wins:

Recognizing and celebrating team achievements is essential for fostering a positive and motivated work environment. As a Scrum Master, I always generated constructive ways to acknowledge my team's accomplishments. This practice of recognizing team achievements served as a clean oil to their engine, which helped to maintain a high team spirit and momentum. Here's an example of how I celebrated a significant team accomplishment:

Scenario 01: I remember when our team successfully delivered a critical Software release. At our monthly team meeting, we invited key stakeholders, including executives and senior leaders, to join us in acknowledging the team's hard work and celebrating the successful completion of the complex project tied with many inter-team dependencies. I prepared a presentation highlighting the project's challenges, breakthroughs, and the exceptional efforts of the team as a whole. We celebrated their dedication, resilience, and collaboration with a small ceremony, presenting each team member with a personalized certificate and a gift card token of appreciation. It was a joyful moment that recognized their hard work and motivated them to continue delivering outstanding results. This event reinforced the team's sense of pride and belonging.

Scenario 02: As our team successfully delivered a complex software release, we organized a celebratory event to acknowledge their hard work and dedication. We rented a nearby venue and held a team appreciation evening. To make it more memorable, I collaborated with the marketing department to create customized trophies for each team member, recognizing their specific contributions to the project. During the event, we shared success stories, laughed at memorable moments, and expressed our gratitude for the collective effort that made the project successful. It was a joy-filled night, a deep sense of belonging and accomplishment.

II. Leaving a Lasting Impact on the Organization:

This section emphasizes the pivotal role of a Scrum Master in enabling long-term organizational transformation. This section delves into the strategies, skills, and mindset required to surpass the initial 90 days and make a lasting impact. It explores how a dedicated Scrum Master can empower teams, facilitate efficient collaboration, and foster continuous improvement, ultimately enabling organizational agility and successful outcomes for projects and the entire organization.

As a long-term Scrum Master, leaving a lasting impact on the organization ensures the continued success and growth of the Agile working methods. Here's an example of how I left a lasting impact:

Scenario 01: As a passionate Scrum Master, I discovered that leaving a lasting impact requires embracing innovation and pushing the boundaries. One of the ways I achieved this was by introducing agile practices beyond the software development realm. I collaborated with teams from different departments, such as marketing and operations, and guided them in adopting agile principles. By breaking down silos and fostering cross-functional collaboration, we witnessed remarkable transformations. Projects became more streamlined, communication improved, and our ability to adapt to change skyrocketed. The ripple effect was astounding as other teams embraced agility, resulting in a culture of innovation and continuous organizational improvement. Witnessing agile practices' positive impact far beyond the software development realm was genuinely inspiring.

Scenario 02: Creating a positive and empowering team culture is essential for leaving a lasting impact. In my previous role, I focused on nurturing a supportive environment where individuals felt valued and motivated. I encourage team members to share their ideas and take ownership of their work. We celebrated successes, learned from failures, and continuously invested in personal and professional growth. This culture of empowerment resulted in a team that was fearless in taking risks and innovating. The energy and enthusiasm were contagious, as team members were inspired to go the extra mile and exceed expectations. Witnessing the transformation of individuals from team members to confident, high-performing professionals

was incredibly rewarding. It served as a reminder that creating a positive team culture can unlock individuals' full potential and drive extraordinary results.

Scenario 03: To foster a continuous improvement culture, I initiated a knowledge-sharing initiative called 'Agile Champions.' We formed a group of enthusiastic individuals from various teams who were passionate about the Agile ways of working. We conducted regular workshops, lunch-and-learns, and brown bag sessions, sharing best practices, emerging trends, and lessons learned. By empowering and inspiring others to embrace Agile, I ensured that the organization would continue to thrive in its Agile evolution long after my time as a Scrum Master in that organization.

Scenario 04: To promote a culture of innovation and continuous improvement, I proposed the establishment of an 'Innovation Challenge.' The challenge encouraged teams to develop creative solutions to existing problems or explore new organizational opportunities. We formed cross-functional teams, provided them with resources and time, and held a grand showcase where they presented their innovative ideas to senior leadership. By sparking enthusiasm for innovation and providing a platform for employees to showcase their talents, we cultivated a continuous learning and exploration culture that would endure long after my tenure as a Scrum Master."

III.   Transitioning Responsibilities with Care:

Transitioning responsibilities as a permanent Scrum Master requires careful planning and effective knowledge transfer. Here's an example of how I transitioned responsibilities with care:

Scenario 01: As I prepared to take on a new role within the organization, I worked closely with my successor, gradually involving them in team meetings and events. We conducted joint retrospectives, where I shared insights, challenges, and strategies for success. I also facilitated shadowing opportunities, allowing my successor to observe my interactions with team members, stakeholders, and external partners. Through open communication and collaboration, I ensured a smooth handover of responsibilities, empowering my successor to continue enabling Agile practices and supporting the team's growth.

Scenario 02: My career path led me to a different department in another organization, so I ensured a smooth transition for my successor. I created a

detailed transition plan, which included one-on-one sessions with my successor to provide insights into the team dynamics, ongoing initiatives, and stakeholder relationships. I also organized a team workshop where we collectively documented our Agile processes, tools, and best practices. Additionally, I encouraged my successor to attend Agile conferences and workshops to enhance their knowledge further. By facilitating a thorough and collaborative handover, I set my successor up for success and ensured a seamless continuation of Agile practices within the team.

By celebrating success, leaving a positive legacy, transitioning responsibilities with care, and extending well wishes, you ensure that the Agile ways of working thrive and the organization continues to reap the benefits of an effective Scrum Master for years to come. Your ongoing dedication and support contribute to a culture of continuous improvement and success.

CONCLUSION

Well, my fellow Scrum Masters, we've ended our adventurous journey together! Can you believe how far we've come? It's been quite a ride from those nerve-wracking first days to becoming confident navigators of Agile success. As we wrap up, let's take a moment to reflect on the key takeaways and lessons we've learned along the way.

Recap of Key Takeaways and Lessons Learned:

Remember the importance of preparing before your start date, the first ninety days, and how it sets the tone for your entire Scrum Master adventure. Be authentic, build rapport, and don't forget to sprinkle a dash of humor to keep things lively. After all, who says work can't be fun?

Embrace the evolution with authenticity and humor. We've discovered that being true to ourselves and injecting humor into our interactions can break down barriers, create positive connections, and foster an enjoyable work atmosphere. So, let your true colors shine, and don't be afraid to laugh!

Inspire others and leave a positive legacy. As Scrum Masters, we have the power to inspire our teams, foster collaboration, and create an environment where innovation thrives. We can make a lasting impact through recognition, celebrating achievements, and a genuine passion for continuous improvement.

And now, my dear fellow Scrum Masters, it's time for us to bid farewell. But fear not! Your Scrum Master's adventure doesn't end here. It's an ongoing evolution of growth, learning, and success. As you continue your path, may you always remember the lessons learned, the relationships built, and the impact you've made.

Wishing You Continuous Success on Your Scrum Master's Adventure!

As I say goodbye, I want to extend my heartfelt wishes for your ongoing success as a Scrum Master. May you navigate the ever-changing waters of Agile with confidence and grace. Embrace challenges as opportunities, inspire those around you, and leave a positive legacy that echoes throughout the organization. Your dedication and passion will continue to make a difference.

Thank you for joining me on this exhilarating Scrum Master's evolution. Sharing my experiences, insights, and laughter with you has been a joy. Remember, you are the ambassadors of change, collaboration facilitators, and Agile excellence champions. Embrace the adventure ahead, and may your Scrum Master's story be filled with achievements, growth, and lasting fulfillment.

ABOUT THE AUTHOR

Karen Fomafung, the founder, and president of BeingAgile Consulting Company is a true visionary and an inspiring leader in the world of Business Agility. With a relentless passion for helping organizations, teams, and individuals unlock their full potential, Karen Fomafung has dedicated 13 years to championing the Agile movement and empowering people to achieve significant value.

With a Master's degree in Agile Product Development and Traditional Project Management, a certified Agile Coach, a certified Scaled Agile Consultant, numerous other Agile certifications, and extensive hands-on experience in Business Agility, she has been shaped as a trusted authority in the Agile space. Known for her unwavering commitment to excellence and ability to inspire positive change, Karen has guided Health, Finance, Tech., and Educational organizations through their adoption of Business Agility, igniting a spark that propels them to new heights.

Her innovative strategies, dynamic training programs, and powerful coaching and mentoring sessions have empowered individuals and teams to embrace Agility at the workplace, foster collaboration, most importantly, help over 7,000 people successfully transition into the Agile career space and surpass their goals.

As a sought-after speaker at international conferences and a respected influencer in the leading digital space, Karen continues to inspire others with her wisdom and expertise. Through her book, Karen shares a step-by-step guide filled with transformative insights, practical techniques, and uplifting anecdotes, guiding Scrum Masters on evolution from day one to extraordinary success.

Prepare to be inspired, empowered, and equipped with the tools you need to explore your full potential as a Scrum Master and make a lasting impact on your organization.

www.ingramcontent.com/pod-product-compliance
Lightning Source LLC
Chambersburg PA
CBHW062248290526
45794CB00006B/2454